FRESHWATER ECOSYSTEMS

by Tammy Gagne

www.12StoryLibrary.com

12-Story Library is an imprint of Bookstaves and Press Room Editions

Produced for 12-Story Library by Red Line Editorial

Photographs ©: Creative Travel Projects/Shutterstock Images, cover, 1; Coy_Creek/Shutterstock Images, 4; Roman Mikhailiuk/Shutterstock Images, 5; Nebojsa Markovic/Shutterstock Images, 6; Baloncici/Shutterstock Images, 7; Erni/Shutterstock Images, 8; risteski goce/Shutterstock Images, 9; Christian Vinces/Shutterstock Images, 10; Vladimir Wrangel/Shutterstock Images, 11, 29 (top left); Ng Zheng Hui/Shutterstock Images, 12; Richard Cavalleri/Shutterstock Images, 13, 15; Helen Filatova/Shutterstock Images, 14; mojoeks/Shutterstock Images, 16; GazTaechin/Shutterstock Images, 17; Pecold/Shutterstock Images, 18; Syndromeda/Shutterstock Images, 19; DM Larson/Shutterstock Images, 20; Katy Foster/Shutterstock Images, 21; Magdevski/iStockphoto, 22; Dudarev Mikhail/Shutterstock Images, 23; Riccardo Mayer/Shutterstock Images, 24; Andrew Zarivny/Shutterstock Images, 25; Pavel L Photo and Video/Shutterstock Images, 26; Syda Productions/Shutterstock Images, 27; Tony Campbell/Shutterstock Images, 28 (top); Strannik_fox/Shutterstock Images, 28 (middle); Vitalii Hulai/Shutterstock Images, 28 (bottom); bancha_photo/Shutterstock Images, 29 (top right); genesisgraphics/iStockphoto, 29 (bottom left); UnicusX/iStockphoto, 29 (bottom right)

Content Consultant: Paul Angermeier, Professor, Department of Fish and Wildlife Conservation, Virginia Tech

Library of Congress Cataloging-in-Publication Data
Names: Gagne, Tammy, author.
Title: Freshwater ecosystems / by Tammy Gagne.
Description: Mankato, MN : 12 Story Library, 2018. | Series: Earth's ecosystems | Audience: Grade 4 to 6. | Includes bibliographical references and index.
Identifiers: LCCN 2016047646 (print) | LCCN 2016051800 (ebook) | ISBN 9781632354563 (hardcover : alk. paper) | ISBN 9781632355225 (pbk. : alk. paper) | ISBN 9781621435747 (hosted e-book)
Subjects: LCSH: Freshwater ecology--Juvenile literature.
Classification: LCC QH541.5.F7 G34 2018 (print) | LCC QH541.5.F7 (ebook) | DDC 577.63--dc23
LC record available at https://lccn.loc.gov/2016047646

Printed in China
022017

Access free, up-to-date content on this topic plus a full digital version of this book. Scan the QR code on page 31 or use your school's login at 12StoryLibrary.com.

Table of Contents

1

Flowing Waters Carry Salt to the Oceans

An ecosystem is a community of living things that share an environment and interact with one another. There are many types of ecosystems. One type is a freshwater ecosystem. It is found in the world's lakes, ponds, rivers, and streams. About 71 percent of Earth's surface is covered with water. But of all the water on the planet, only 3 percent is freshwater. The rest is salty ocean water. Freshwater is the only type of water that people can drink.

Flowing waters sculpt the landscapes we see and live in. They include rivers and streams that constantly move in one direction. This movement is called the current. It carries the water downhill toward another body of water. One river or stream may lead to another. Eventually, flowing waterways empty into the ocean.

Salt is found in the soil that surrounds rivers, streams, and other waterways. Flowing freshwater

Freshwater ecosystems are also called lotic ecosystems.

carries salt from the land to the sea. Rivers and streams constantly pour salt into seawater. But water from a river or stream does not taste salty like ocean water does. This is because the ratio of water to salt is so high in freshwater.

Rivers and streams are a big part of freshwater ecosystems. They support many types of animal and plant life. Plants provide food for river animals. Smaller animals become food for larger or more powerful animals. Every living thing in an ecosystem is connected.

THINK ABOUT IT

Based on the information you have read here, do you think most rivers begin high in the mountains or at sea level? Why?

4 billion

Amount, in tons (3.6 billion metric t), of dissolved salt that rivers carry into the world's oceans each year.

- Rivers and streams move downhill toward the ocean.
- Flowing waters carry salt from the land to the sea.
- Freshwater ecosystems support a wide variety of animal and plant life.

2

The Nile Is the World's Longest River

The Nile River is approximately 4,132 miles (6,650 km) long. It is the longest river in the world. The exact location of its source is unclear. Many people think the river begins in Burundi. This is a tiny African nation just south of the equator. Others think the Nile's source is in Rwanda. The river flows north all the way to Egypt. There, it empties into the Mediterranean Sea.

The Nile ecosystem is filled with many life-forms. Lotus flowers grow in shallow areas. Papyrus plants grow in marshy areas. The river is

The Nile has many smaller rivers and streams flowing into it.

GOING TO WAR FOR WATER

In 2011, Ethiopia began building a dam that would block much of the Nile's water from reaching Egypt. Both countries needed the river for drinking water and growing crops. The Egyptian government said that it would go to war with Ethiopia if it went through with the project. Without the Nile's water, the Egyptian people would suffer.

10

Number of countries that share the Nile River.

- The Nile is the longest river on Earth.
- People use the Nile's freshwater for drinking, fishing, and farming.
- The silt at the Nile Delta is good for farming.

also home to many types of fish. Nile perch, catfish, and tilapia are common. The Nile crocodile's diet is made up largely of these fish.

Much of northeast Africa has a hot climate and a lack of water. Many countries depend on the Nile. It is the main source of drinking water for all of the nations the river runs through. People also use the Nile's freshwater for fishing, transportation, and watering crops.

More than half of Egyptians live in the Nile Delta.

The Nile Delta is a triangle-shaped piece of land near the mouth of the Nile in Egypt. It is here that the Nile River meets the sea. The river deposits large amounts of silt in this region. The water and silt make the Delta one of the best areas in Africa for farming. Silt holds in moisture and nutrients better than other types of soil.

3

Many Animals Live in Freshwater Ecosystems

Freshwater is packed with many life-forms. Some animals, such as fish and turtles, make their homes in the water. Other animals, including many birds, live near freshwater ecosystems. All of these animals depend on freshwater for food.

Loons are birds that are common to freshwaters. They nest and breed in these ecosystems. They eat fish, insects, and even frogs that they find in freshwater. Ducks and geese eat the leaves, seeds, and stems of freshwater plants.

Freshwater animals can be as diverse as the waterways

SIGNS OF HEALTHY ECOSYSTEMS

Dragonflies are found near many rivers, streams, ponds, and lakes. As larvae, dragonflies spend most of their lives underwater. They depend on clean water for survival. A large number of dragonflies in an area suggests that the water, plants, and other animals in the ecosystem are also thriving.

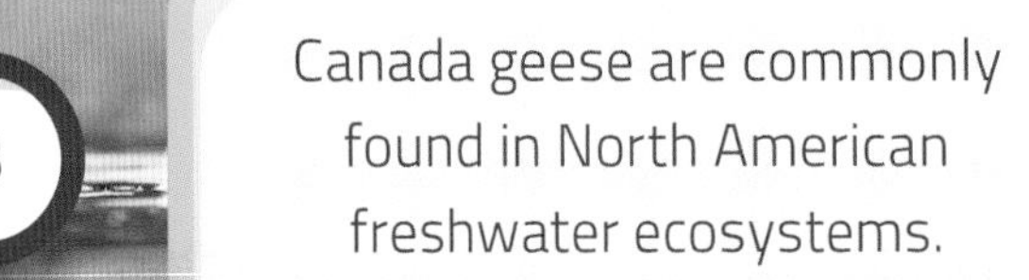

Canada geese are commonly found in North American freshwater ecosystems.

Piranhas are native to South American freshwaters.

themselves. The anaconda is the world's largest snake. It spends much of its time in the freshwater ecosystems of South America. The diving bell spider is the only spider species that lives underwater. It can be found in the freshwaters of Europe. Mammals also live in many freshwater ecosystems. The North American river otter is found in both rivers and lakes. Beavers are common to many freshwaters, too.

The blue catfish is found in the freshwaters of Ohio, Missouri, and Mississippi. It can grow up to five feet (1.5 m) long. It weighs as much as 100 pounds (45 kg). It eats any other fish it finds. The piranha is native to the Amazon River in South America. It measures just 12 inches (30 cm) long. It is known for its razor-sharp teeth. Piranhas mainly eat smaller fish. But they are aggressive predators. They will eat other piranhas when food is scarce.

10,000

Number of fish species that live in the world's freshwaters.

- A wide variety of animals live in freshwater ecosystems.
- Some animals eat plants that grow in freshwaters.
- Some freshwater animals survive by eating other animals in the ecosystem.

4

The Amazon River Flows through a Rain Forest

The Amazon rain forest in South America is home to more than 40,000 known plant species. It also has 3,000 fish species and 1,300 bird species. But scientists believe there are many more plants and animals yet to be discovered.

The Amazon River runs through the Amazon rain forest. The river is slightly shorter than the Nile. It is approximately 4,000 miles (6,400 km) long. That is roughly the distance between New York City and Rome, Italy. But the Amazon is wider than the Nile. It is approximately 6.8 miles (11 km) wide at its widest point, even during the dry season. In the wet season, the Amazon can grow to 24.8 miles (40 km) at its widest point.

The Amazon River empties into the Atlantic Ocean in northeastern Brazil. Five different tributaries could be its source. Some scientists

The Amazon carries the largest volume of freshwater of any river in the world.

The South American tapir is one of many animals that rely on the Amazon River's freshwater for survival.

think the Amazon begins at the Mantaro River in Peru. If they are right, it means that the Amazon is up to 57 miles (92 km) longer than originally thought.

Many people who live near the Amazon River have little to no contact with the rest of the world. These indigenous peoples have lived in the rain forest for many generations. They know very little about life beyond the Amazon.

1,100

Number of tributaries that flow into the Amazon River.

- The Amazon River is part of a large and important ecosystem.
- The Amazon River is the widest river in the world and the second longest.
- The Amazon's waters and banks are home to many plants and animals.

5 Still Waters Have Little Current

Unlike rivers and streams, still waters have little current. Still bodies of water include lakes and ponds. They have only a small amount of flow into or away from them. Most still waters are surrounded by land. To maintain their water levels, still waters need regular rainfall or melting snow.

Scientists do not have a standard method for measuring the size of lakes and ponds. Some people measure the surface area. Others measure the depth. This means there is no official way to tell the difference between lakes and ponds.

Most lakes are much larger and deeper than ponds. The deepest lake in the world is Siberia's Lake Baikal. It measures 5,354 feet (1,632 m) at its

Blue Pond is located in Biei, Hokkaido, Japan.

395

Length, in miles (636 km), of Lake Baikal in Siberia.

- Still bodies of water do not flow as rivers or streams do.
- Sunlight can typically reach the bottom of a pond to maintain plant life.
- Lakes are usually bigger and deeper than ponds.

Echo Lake in Conway, New Hampshire, is just 14 acres (5.7 ha) in area and 11 feet (3.4 m) deep.

deepest point. It holds more freshwater than any other lake on Earth.

Many plants found in ponds and lakes have roots that are buried in the mud. It is hard for them to get the oxygen they need to survive. Instead, they take in oxygen through their leaves. The oxygen is then transported to the plant's roots.

Bacteria, fungi, insects, worms, plankton, and snails are common organisms found in lakes and ponds. These freshwater ecosystems are also home to many types of fish, including catfish, largemouth bass, and minnows. Birds that call these freshwater ecosystems home include kingfishers, mallards, and herons. Toads, newts, and salamanders are some of the amphibians that live in ponds and lakes. You can even find mammals such as beavers, muskrats, moles, and raccoons.

THINK ABOUT IT

Based on the information in this book, do you think scientists could find a way to distinguish lakes from ponds? How might they do this?

6

A Melting Glacier Formed the Great Lakes

The Great Lakes make up the largest group of freshwater lakes in the world. The five lakes are located between the upper Midwest region of the United States and southern Canada. They include Lake Erie, Lake Huron, Lake Michigan, Lake Ontario, and Lake Superior.

These huge bodies of water formed more than 20,000 years ago at the end of the last Ice Age. A massive glacier melted, carving craters into the land below it. The meltwater from

Pukaskwa National Park in Ontario, Canada, sits on Lake Superior.

95,160

Surface area, in square miles (246,463 sq km), of the Great Lakes.

- The Great Lakes are located along the border of the United States and Canada.
- These five freshwater lakes formed from a melting glacier more than 20,000 years ago.
- People in both countries rely on these large bodies of water in many ways.

Chicago, Illinois, was built alongside Lake Michigan.

the glacier filled the craters with freshwater.

Many animal species live in or near the Great Lakes. Gray wolves and Canada lynx fish and drink from these freshwaters. The types of fish found in the Great Lakes vary based on temperature. Yellow perch and white bass live in the warmer waters of Lake Erie. Brook trout prefer the colder waters of Lake Superior.

People in both the United States and Canada use the water from the Great Lakes. The lakes provide more than 40 million people with drinking water. People also use 56 billion gallons (212 billion L) of the lakes' water each day for different jobs. Fishing is common in the Great Lakes. Lake trout serve as food for many animals within the ecosystem. People also make a living catching and selling fish.

ONE LAKE LEADS TO ANOTHER

All five Great Lakes are connected to at least one of the other Great Lakes by a river or strait. A strait is channel of water connecting two larger bodies of water. St. Mary's River connects Lake Huron to Lake Superior. The Niagara River links Lake Erie to Lake Ontario. Lake Michigan and Lake Huron are joined by a narrow waterway called the Straits of Mackinac.

7

Plants Help Keep Freshwater Healthy

Most plants need freshwater for their survival. But plants also serve important purposes for their freshwater ecosystems. Animals, such as moose and otters, rely on freshwater plants for food. Birds and fish use plants as safe places to lay eggs or hide from predators. Plants even keep freshwater clear by using nutrients that would lead to the growth of algae. A small amount of algae is healthy. But too much algae is a sign of an unhealthy ecosystem.

Among the most helpful freshwater plants are cattails. These tall plants trap pollutants from runoff water. This is water from rainfall and melting snow that travels across the ground and into waterways, carrying pollutants with it. Cattails also protect the shoreline when storms occur. Wind and rain can destroy land where it meets the water. Cattails help absorb the storm's energy before it reaches the land.

Otters are commonly found in freshwater ecosystems.

60

Number of orchid species that grow along Lake Superior's shoreline.

- Freshwater plants serve as food, hiding places, and nesting grounds for many animals.
- They also help keep water clear by using up nutrients that lead to the growth of algae.
- Cattails protect both freshwater and its shorelines.
- Many animals eat freshwater plants, such as water lilies and watercress.

Cattails act as natural filters for freshwater ecosystems.

Another common freshwater plant is the water lily. More than 70 species grow throughout the world in a range of water temperatures. Lilies have roots that attach to the waterway's floor. They have flowers and leaves that float on the water's surface. Turtles, fish, and snails eat the leaves.

Watercress is also found in many freshwaters. Like water lilies, watercress flowers and leaves float on the surface of freshwater. Ducks, muskrats, and deer enjoy eating the leaves of this cabbage-like plant.

8

Lake Victoria Is Africa's Largest Lake

A British explorer named John Hanning Speke came upon one possible source of the Nile River in 1858. He named the large body of water Lake Victoria after the United Kingdom's ruling queen. Measuring 26,828 square miles (69,484 sq km), Lake Victoria is located in Tanzania, Uganda, and Kenya.

One of the most common plants in Lake Victoria is the water hyacinth. But this beautiful flower is an invasive species. It has caused problems in the Lake Victoria ecosystem. Too many water hyacinths cover large areas of the lake's surface. They take up

Lake Victoria is second in surface area only to Lake Superior in the United States.

2,138

Length, in miles (3,441 km), of Lake Victoria's shoreline.

- Lake Victoria is one possible source of the Nile River.
- It is located in three nations—Kenya, Tanzania, and Uganda.
- The lake is used for fishing, travel, and trade between the three countries.
- Pollution is putting the health and future of Lake Victoria at risk.

Pollution threatens Lake Victoria's fishing communities.

TROUBLE WITH TILAPIA

More than 200 fish species live in Lake Victoria. Tilapia is the most important fish for the local economy. But the fish are not as plentiful as they once were. More than 2 million people make their living fishing in this region. Overfishing and pollution have caused tilapia populations to decrease. As a result, the price of tilapia has doubled in recent years. Over time, people may not be able to support themselves by fishing for this species.

space other plants and animals need. Using chemicals to kill water hyacinth could hurt other living things in the lake. Experts have brought South American weevils to the region to help eat the water hyacinths. These beetles eat large amounts of the plant to help keep it from taking over other plants.

Lake Victoria serves many purposes for the people who live near it. Many people make their living by fishing this lake. They also use the lake to travel between the three countries it borders. Many companies use the lake for trade between these three nations.

Lake Victoria provides drinking water to people in the surrounding areas. But pollution has become a problem for the lake in recent years. Human waste and chemicals have made the water unsafe in some areas. Too much algae growth has lowered the lake's water quality. The toxins from some types of algae can harm both animals and people.

9

Wetlands Protect Other Ecosystems

Freshwater wetlands are areas where land is covered by shallow water. These ecosystems sometimes overlap other ecosystems. Wetlands are found at the edges of large lakes, ponds, and rivers. Some wetland areas change with the seasons. They may flood during certain times of year. They may remain dry at other times of year. Rainy seasons often bring too much water. Hot, dry periods can cause some water to evaporate.

Wetland animals and plants have adapted to live in places where the water rises and lowers. Many animals move in and out of

THINK ABOUT IT

Based on what you have read about wetlands, do you think it is important to protect them? What might happen if the wetlands were to disappear?

A wetlands area near Mackay, Idaho

50

Percentage of the United States' original wetlands that remain today.

- Wetlands are covered by shallow water that rises and falls over time.
- Plants and animals must be able to tolerate water-level changes.
- Wetlands protect nearby ecosystems, both on the land and in water.

wetlands. Several bird species fly to wetter regions during dry seasons. Other animals remain in the wetlands all the time. Frogs live in the wetlands year-round. The African lungfish lives in the small rivers of West and South Africa. It survives dry periods by burying itself in the mud for months. It remains underground inside a slimy cocoon. This keeps the fish wet despite the dry conditions.

Pollutants are often trapped by wetland plants and sediment.

Wetlands perform important jobs for local ecosystems. They provide a breeding ground for insects, such as mosquitoes. Birds and fish eat the insects. Some wetlands' location between the mainland and the open sea allows them to shield the mainland from storms. They also trap pollution before it can enter nearby waterways. This helps keep pollutants from moving into rivers and other bodies of water. Wetlands are filters that protect freshwater resources.

10

Humans Affect Freshwater Ecosystems

Many human activities can hurt freshwater ecosystems. Pollution is one of the biggest problems. Dumping wastewater into freshwater is a huge concern. Some people also throw garbage into waterways. Rainwater can carry garbage and chemicals into freshwater. When this happens, water quality declines. Animals that need the water to drink and as a place to live also suffer. Some animals move to other areas. Others become sick or die.

780 million

Number of people who do not have access to clean drinking water around the world.

- Pollution is among the biggest threats to freshwater ecosystems.
- Overfishing can make it hard for species to maintain their populations.
- Nonnative species can harm the animals and plants within a freshwater ecosystem.

Trash can be harmful to freshwater animals.

When people overfish freshwater ecosystems, fish populations decline.

USING TOO MUCH WATER

A drought is a long period of time without rain. It can lead to a shortage of water in hot, dry places. When a drought occurs, it is vital for people not to waste water. People can avoid using water for things that are not necessary, such as watering lawns. They can also not let tap water run while brushing their teeth or washing dishes. Without water, farmers cannot grow crops. In some areas, droughts last so long that they threaten access to safe drinking water.

Another way people hurt freshwater ecosystems is by overfishing. When too many fish are taken from freshwaters, the ones that remain often have a hard time maintaining the population. There are not enough of them to reproduce. Some species may become extinct.

Some people hurt freshwater ecosystems by bringing invasive species into them. Nonnative species can include both animals and plants. An invasive plant called hydrilla is a big problem in Florida. It grows as dense mats, blocking sunlight for other plants. It also provides a poor habitat for fish. European carp are an invasive fish species causing problems in Tasmanian lakes. The carp eat food that other fish in this ecosystem need. They also threaten the populations of smaller fish by eating too many of them.

11

People Depend on Freshwater Ecosystems

People need clean freshwater every day. The human body needs between 9 and 13 cups (2.1 to 3.1 L) of water each day to remain healthy. Water helps carry nutrients through a person's body. It also helps remove waste in the form of urine. Without water, the body becomes dehydrated. This is a condition that makes a person feel tired and sick.

People use water to grow food. A crop cannot grow without it. Fruits, vegetables, and grains all need water to grow into edible foods. Nearly all foods contain water. Even dry foods, such as bread and cereal, contain at least some water.

Over time, people have learned how to use water for helpful purposes. Without water, people would not have flushable toilets, dishwashers, or washing machines. Scientists have even learned how to use water to produce electricity.

The Hoover Dam on the Colorado River uses water to spin

People need clean drinking water to survive.

The Hoover Dam is just one of many plants in the world that produce electricity from water.

large blades called turbines. The pressure of the water coming from the dam turns the turbines. This movement creates energy in the form of electricity. This type of energy is called hydroelectric power. It is a renewable resource. This means it does not run out. The water just keeps flowing and creating more energy. The electricity can then be transmitted to millions of people. They can use it to light and heat their homes and other buildings.

3

Number of days a person can survive without drinking water.

- People become tired and sick if they do not drink enough water.
- Freshwater makes it possible for people to grow food.
- Freshwater is used to make hydroelectric power.

LOCKED AWAY

More than two-thirds of all the freshwater on Earth is not available to people. It is frozen in glaciers at the North and South Poles. These glaciers reflect the sun's rays. They help keep the planet from becoming too hot. But many glaciers are melting due to global warming.

12

People Can Help Care for Freshwater Ecosystems

Many freshwater ecosystems are in trouble. They need people to care for them.

One of the easiest ways to help freshwater ecosystems is by keeping pollution out of them. This can be done by putting trash in its proper place. People can help throw away trash that others have left behind. When everyone lends a hand, people can make a big difference.

People can help freshwater ecosystems by cleaning up the trash that gets into them.

14

Percentage of indoor water use that is lost to leaks each day.

- Preventing pollution helps protect freshwaters.
- Reducing water usage prevents water shortages.
- Eating only sustainable fish can help reduce overfishing.
- Conservation organizations can help people learn how to protect freshwater ecosystems.

One way to help freshwater ecosystems is by telling others how they can help protect them.

Conservation is another important step when it comes to caring for freshwaters. When people use only the water they need, they reduce waste. One way to reduce waste is by turning off the faucet when not in use. Another way is to collect freshwater in rain barrels to use for watering gardens.

People who do not fish can help reduce overfishing by making smart choices when shopping. A big reason for overfishing is high demand. Learn about the types of fish that are the most sustainable and eat only those species.

One of the best ways to help freshwater ecosystems is by sharing what you learn about them. Tell others about the importance of freshwater ecosystems. Encourage them to get involved in caring for these ecosystems. Conservation International and the Sierra Club are just two of the many organizations that work to protect freshwater ecosystems around the world. People can help by donating to one of these groups or visiting their websites to learn more about conservation.

Freshwater Food Web

northern pike
snails
yellow perch
algae

Glossary

algae
A plantlike organism that grows in water.

current
The part of a body of water that is constantly moving in one direction.

evaporate
To turn from a liquid into a gas.

glacier
A large body of ice found in the northernmost or southernmost parts of the planet.

invasive
Tending to spread and invade healthy ecosystems.

pollutants
Harmful materials that spoil the environment.

silt
Small material carried by a river that settles on the bottom or near the mouth of a waterway.

sustainable
The ability of a species to thrive despite regular harvesting.

tributaries
Streams or small rivers that flow into larger rivers.

For More Information

Books

Conrad, Steve. *Enough Water? A Guide to What We Have and How We Use It*. Ontario, Canada: Firefly, 2016.

Gagne, Tammy. *Rain Forest Ecosystems*. Minneapolis, MN: Abdo Publishing, 2016.

Newland, Sonya. *Journey along the Nile*. London, UK: Wayland, 2016.

Visit 12StoryLibrary.com

Scan the code or use your school's login at **12StoryLibrary.com** for recent updates about this topic and a full digital version of this book. Enjoy free access to:

- **Digital ebook**
- **Breaking news updates**
- **Live content feeds**
- **Videos, interactive maps, and graphics**
- **Additional web resources**

Note to educators: Visit 12StoryLibrary.com/register to sign up for free premium website access. Enjoy live content plus a full digital version of every 12-Story Library book you own for every student at your school.

Index

About the Author
Tammy Gagne has written more than 150 books for both adults and children. She resides in northern New England with her husband and son. One of her favorite pastimes is visiting schools to talk to children about the writing process.